ABANDONED AFTER THE STORM

HURRICANE MICHAEL

KIM HILL

AMERICA THROUGH TIME®
ADDING COLOR TO AMERICAN HISTORY

America Through Time is an imprint of Fonthill Media LLC
www.through-time.com
office@through-time.com

Published by Arcadia Publishing by arrangement with Fonthill Media LLC
For all general information, please contact Arcadia Publishing:
Telephone: 843-853-2070
Fax: 843-853-0044
E-mail: sales@arcadiapublishing.com
For customer service and orders:
Toll-Free 1-888-313-2665

www.arcadiapublishing.com

First published 2020

ISBN 978-1-63499-221-3

Typeset in Trade Gothic 10pt on 15pt
Printed and bound in England

CONTENTS

ABOUT THE AUTHOR

KIM HILL is a freelance writer, website designer, painter, award-winning photographer, and overall Jill-of-all-trades who lives in sunny, and often stormy, Florida—where there's always something happening in the news related to the state. She has two degrees in criminology and is an advocate for SkinCancer.net, where she helps educate others on what to expect when dealing with skin cancer.

Kim recently had three of her photographs selected for the Clio Art Fair in Chelsea, NYC. She has contributed to *In the Eye of the Storm: Stories of Survival and Hope from the Florida Panhandle, Survivors: Work Created in the Wake of Hurricane Michael, In the Aftermath of the Storm: Stories of Hope and Healing, In Laymon's Terms*, and *Abandoned Northwest Florida*. She also has a horror author pen name she writes under.

When she's not working, Kim enjoys road trips, movies, and incessant reading way past her bedtime.

INTRODUCTION

On October 10, 2018, the Florida Panhandle took a direct hit by Hurricane Michael. It was the third most intense storm in terms of pressure and the strongest in terms of wind since. Nothing would match it since Hurricane Andrew. This storm cut a swathe through Central America and Cuba before targeting the Gulf Coast. It devastated parts of Florida but wasn't finished. It then moved on through Georgia, the Carolinas, and Virginia. Even Maryland saw remnants of the storm—as well as Europe.

After Hurricane Michael's destruction, homes and businesses lay in pieces. Ranging from damaged to destroyed, many now stand abandoned, with their residents or business owners left to decide whether to repair, rebuild, or move on.

This is their story. A glimpse at the abandoned places that were homes, businesses, and transportation to many. A year after the category 5 storm, there are still buildings that look exactly as they did the day after the hurricane. There is a caveat to all this that some of these places may no longer be standing, especially in Mexico Beach. Yet many of them look exactly like you will see in the pictures within this book. In fact, some of the photos were taken as late as September of 2019, almost a year later.

A house moved into the water

A STORM IS BREWING

Hurricane Michael started out as most do—a little atmospheric phenomenon that no one worried about much. It began as a disturbance on October 1, then became a tropical depression on the 7th. But still, it was nothing to worry over. By the 8th, it intensified into a hurricane near the tip of Cuba. Yet, it was still a smaller hurricane that did not cause seasoned Floridians to fret.

Within twenty-four hours, it went from a category 1 storm to a category 4, and was later reclassified as a category 5. It was too late for those who stayed to leave the area.

Hurricane Michael made landfall near the tiny coastal town of Mexico Beach. It was so strong that it not only destroyed much of Mexico Beach, it also destroyed parts of Panama City and its surrounding townships to the west.

As it moved inland, the storm lessened and began to take a northeastward trajectory toward Chesapeake Bay, weakening to a tropical storm over the state of Georgia. It then transitioned into what is known as an extra-tropical cyclone and made its way over southern Virginia on October 11. Michael subsequently strengthened into a powerful extra-tropical cyclone and eventually made its way all the way to Europe, where it impacted the Iberian Peninsula before diminishing on October 16.

Five states declared states of emergency due to Hurricane Michael: Florida, Alabama, Georgia, South Carolina, and North Carolina. Each declared emergencies in some or all their counties.

A COSTLY STORM

According to sources:

> Hurricane Michael caused an estimated $25.1 billion (2018 USD) in damages, including $100 million in economic losses in Central America, damage to U.S. fighter jets with a replacement cost of approximately $6 billion at Tyndall Air Force Base, and at least $6.23 billion in insurance claims in the U.S.[2]

While the people of the Panhandle and surrounding areas are picking up the pieces, thousands of buildings were destroyed or left abandoned. This is the story of those buildings. Their beauty lies in their strength of weathering the storm—or at least trying to.

1

MEXICO BEACH

The hardest hit area by Hurricane Michael was Mexico Beach. Mexico Beach is a small town referred to as "Mayberry on the Water." It's just east of Panama City and has only a little over 1,000 people who live there permanently. It's a friendly town where people say hello as you pass by. Pastel colored buildings dotted the landscape, and it was simply a walk across the road, or sometimes out your front door, to get to the beach.

Whether it's fishing, swimming, or simply enjoying the view—Mexico Beach attracted tourists, but was rarely crowded. Sure, major holidays brought in the visitors, but this was still an area unspoiled by looming condominiums or big box stores. Instead, there were mom-and-pop shops, local restaurants, and places to stay that were family-owned businesses.

When Hurricane Michael made landfall, it caused an up to 20 foot storm surge that swept away buildings, boats, and pretty much anything else standing in its way. In fact, the United States Geological Survey (USGS) found high-quality water marks in three areas that showed between 17 and 19 feet of water above mean sea level (MSL). These were found in buildings where waves could not reach, which is why they are considered an accurate depiction. When you combine the storm surge along with waves, the high-water mark (HWM) measured in at 20.6' above mean sea level. Just a few miles away in Port St. Joe, a high-water mark measured even higher at 21.2'. This data makes it the sixth highest in the past eighty-five years, with Hurricane Katrina being in first place at 34.11'.

[*Photo courtesy of Bill Fauth*]

A house in Mexico Beach

[*Photo courtesy of Bill Fauth*]

[*Photo courtesy of Bill Fauth*]

[*Photo courtesy of Bill Fauth*]

[*Photo courtesy of Bill Fauth*]

Boat storage photo [*courtesy of Bill Fauth*]

[*Photo courtesy of Josh Pyfrom*]

[Photo courtesy of Josh Pyfrom]

THE DRIFTWOOD

From homes and businesses to cars and boats, Mexico Beach took the ground zero hit from Hurricane Michael. One of the destroyed buildings is a place known as Driftwood Inn. For four generations and over forty-three years, the family-run motel had seen its share of celebrations, visitors, and even a few storms. It stood all that time until the hurricane swept through, tearing holes throughout the building and making it unlivable.

> Florida's building code, put into effect in 2002, is famously stringent when it comes to windstorm resistance for homes built along the hurricane-prone Atlantic shoreline. But it is less so for structures along the Panhandle, a region historically unaffected by storms as strong as the ones that have slammed into South Florida.
>
> After Hurricane Andrew, a Category 5 beast, ravaged Miami-Dade County in 1992, new construction in the southern portion of the state was required to withstand 175-mile-an-hour winds. In the coastal Panhandle counties affected by Michael, the requirement is lower, for 120 to 150 miles an hour, and the rules for certain kinds of reinforcement have applied to houses built more than a mile from shore only since 2007. Many of the residences and businesses rubbed out by Michael in Mexico Beach were far older; rebuilding them to conform to the new code will be expensive, and could price out some of the working-class people who historically have flocked to Mexico Beach.[1]

Driftwood Inn [*Photo courtesy of Bill Fauth*]

Driftwood Inn [*Photo courtesy of Bill Fauth*]

Hope on the horizon [*Photo courtesy of Bill Fauth*]

2

CHURCHES

Some might think that it would seem that a church would be the safest place to be in a storm—or at the very least, a fairly safe shelter. But some of the buildings hardest hit by Hurricane Michael were churches. This is most likely due to their very structure with tall walls and large open rafters. The wall of a church, which has much less internal support in the higher levels, just can't take the pressure. And Hurricane Michael packed a lot of pressure—a minimum central pressure of 919 millibars.

Another reason many churches may have been damaged is due to location. Larger churches often sit on large patches of land with open areas devoid of trees. These open areas gave the hurricane a direct path with not much protection from other buildings or trees. And while some buildings took major damage from trees being felled, in some cases it was the trees that helped protect the buildings.

In fact, this church in the Pretty Bayou area looked as if a tornado hit it. According to local weather, there were no tornadoes in Panama City while Michael spawned sixteen tornadoes from Florida to Virginia overall. What may have happened to this church is what is called a microburst. Both tropical storms and hurricanes can create these wind bursts that are highly dangerous. A microburst is a localized column of sinking air (downdraft) within a thunderstorm and is usually less than or equal to 2.5 miles in diameter. They can reach speeds of up to around 100mph and combined with the winds of a hurricane, you have a very dangerous situation that can mimic the damages of a tornado.

And while it is unlikely that a microburst affected all of the churches in the area, this one looks to have damage outside of the hurricane's power.

These are just a sample of the damages done to the area's churches and this is in Panama City, alone. There are many more that look just as damaged, if not more so.

One of the many damaged churches in Panama City and surrounding areas

Family of God Baptist Church

Family of God Baptist Church

Family of God Baptist Church

Damaged church

Extreme damage to the church

Damaged church in Callaway

The same damaged church in Callaway

A rainbow of hope

Presbyterian Church

Presbyterian Church

Presbyterian Church

Presbyterian Church Educational Center

Presbyterian Church

Presbyterian Church Educational Center

Presbyterian Church

Above: Presbyterian Church

Left: Church steeple

3

MARINAS AND BOATS

It's no surprise that the Panhandle's marinas and boats took a huge hit. Destroyed were entire marinas with millions of dollars' worth of boats. People's livelihoods, retirement, and even homes got tossed around in the wind.

Many of the marina areas have been cleaned up or are in the process of being cleaned up. But in some areas, boats still sit on land, as they did a year ago. Those are possibly abandoned or perhaps their owners do not have the funds to get them yet.

And while the marina in Mexico Beach will reopen, the Panama City Marina closed indefinitely after standing more than fifty-nine years. The insurance companies are working with owners to have their boats removed but it closed to foot traffic due to safety concerns. The damage was too severe for the marina to function.

Boat in Mexico Beach [*Photo courtesy of Bill Fauth*]

Massalina Bayou

Massalina Bayou

Massalina Bayou

Massalina Bayou

Massalina Bayou

Above: A grounded boat
Below: Another grounded boat
Inset: Grounded boat at Carl Gray Park

Grounded boat in the Cove

Damaged boat behind Rodeo's Restaurant in Callaway

Panama City Marina

Panama City Marina

A ship is righted

Massalina Bayou

4

SURROUNDING HOMES

Pick a spot anywhere in the Hurricane Michael hit area of the Panhandle and you won't have to go too far to find an abandoned home. In fact, abandoned homes alone could fill an entire book, but there are too many to find or photograph. Blue tarps are a common sight but many homes sit empty, devoid of any repairs.

This is due to a few reasons. In some cases, the person living in the home may have moved away. The building may be too expensive to repair or is not repairable at all. Some are waiting on insurance claims even a year later. And some are simply waiting to be fixed for a new family to move in. Yet most of them in this book are probably going to sit exactly as they were for whatever reason.

Many have tree damage or structural damage that cannot be fixed for what the house is worth. Unfortunately, this has caused the cost of housing to go up in the area and real estate is at a prime. Especially when you have entire apartment complexes that were destroyed.

Public domain photo

[*Photo courtesy of Bruce Benedict*]

[*Photo courtesy of Mike Cassini*]

Home with damage

Apartment complex in Lynn Haven

Apartment complex in Lynn Haven

Apartment complex in Lynn Haven

THE DOLLHOUSE

This house has been referred to as the "Dollhouse." This is because the hurricane sheared off the entire side, leaving furniture as it was and making it look like a dollhouse. This happened to a few houses where entire walls were torn away leaving furniture, appliances, and ceiling hanging from a mildewed ceiling.

These are also a sample of the thousands of damaged homes in the area, especially in Lynn Haven and Callaway—and of course, Mexico Beach.

Above: Dollhouse [*Photo courtesy of Chastity Wofford*]

Left: Dollhouse photo

Dollhouse photo

Dollhouse photo

Dollhouse photo

Dollhouse photo

Home in Panama City

Back view of the home

Home in Callaway

Damaged home

A home cut in half

Damaged home

Damaged home

Home in Lynn Haven

Damaged home

Damaged home

Damaged home

Home in Mexico Beach that was moved into the water

5

BUSINESSES, SURROUNDING BUILDINGS, AND SCENERY

Just like everything else, businesses took a hard hit from the storm. Some were totally destroyed and some are waiting to be repaired, torn down, or left abandoned. Here are just a few of the many businesses, surrounding buildings, and scenery from the post-Michael landscape in the Panhandle of Florida.

Every day it seems that a new place is back in business but with every one that has returned, there seems to be two more that are still just sitting.

Both of Panama City's movie theaters were damaged. Both theaters were located on the main strip of 23rd Street. The Carmike Theater and the Regency Theater were damaged beyond repair. So far, the Regency has been torn down in order to be rebuilt and the Carmike's fate is still up in the air.

Grocery Outlet [*Photo courtesy of Laura Risner*]

Downtown Panama City

Downtown Panama City

Downtown Panama City

Downtown Panama City

Regency Theater

Carmike Theater

Carmike Theater

All types of businesses were damaged.

Hwy 98 near Harrison Ave

Downtown Panama City

Downtown

Damaged garage

Damaged building

Damaged office

Damaged garage

Building complex

Building complex

More graffiti at a building in Callaway

Luigi B. G.

Humane Society Thrift Store

Damaged nail place

Inside of a damaged building

Building complex

More of the damaged building complex

Core & Main

More Core & Main

Damaged offices

Damaged warehouse

Rodeo's Restaurant

Rodeo's Restaurant

Damaged office on 23rd St

Damaged office off of Harrison

Wiped out building

This motel could not be salvaged

Destroyed motel in Callaway

Damaged beyond repair

Damaged motel

Damaged Wendy's

Building in Callaway

Old Ollie's building in Callaway

Damaged building

Destroyed building in front of a working Subway restaurant

Destroyed apartment building

Beyond repair

The roof created a mosaic of destruction

The tarp doesn't seem to have helped this building

Building in front of the destroyed Ollie's store

Damaged building

Old motel that could not be saved

Inside the motel office

Motel office

Oriental Inn Restaurant

This entire strip mall was destroyed

Destroyed strip mall

Damaged building

Destroyed Metro PCS

Former nail place

Former jeweler

6

TIME WILL TELL

Hurricane Michael destroyed lives, livelihoods, and buildings. New buildings are going up and some places are being repaired and reopened. But Hurricane Michael changed the landscape of the Panhandle of Florida.

Train boxcars were blown over like toys, yet they were picked up and re-positioned. Trees fell by the thousands and many remain where they toppled. Buildings were damaged severely, and some remain as they were while others are repaired or torn down. The hurricane created an entire area filled with abandoned homes, businesses, and transportation. And unless another storm of equal power comes around, some may stand for another 100 years—empty of humans, with only memories left inside.

Damaged building

What used to be a public park

Asbell Park

IN CLOSING

Here is something I wrote right after the storm. If you do not live in this area, maybe it'll give you a sense of how it felt during and immediately after. Maybe it will help you relate to the people whose damaged lives fill the pages.

I'm not even sure where to begin when it comes to Hurricane Michael. Everything is still so fresh and raw, and there are so many people who lost everything. All I can do is tell it from my experience, because that's the only way I know how to express it.

I have lived on the Gulf Coast my entire life. Often, people wonder why we live in an area prone to hurricanes. Well, let me tell you a little about the Gulf Coast of Florida.

We have some of the most beautiful beaches in the world, complete with miles of sugar-white sand, emerald waters, and a diverse mix of marine life. From dolphins and sharks to manatees and sea turtles—I've seen them all. The skies are sunny and blue, although in a second that can change; the heavens will turn dark and the rain will pour. To me, that's the most beautiful of all.

The Gulf Coast has a mix of people as diverse as its fauna, flora, and marine life. There are people here that have spent their entire lives in one place, there are millions of tourists from all over the world who descend upon the area, and we're home to a variety of celebrities—from Emeril Legasse to Anne Rice—not to mention those who just enjoy visiting. But we're not Hollywood. We're hospitable and friendly, and you'll find the best stories from a diner waitress, an oyster shucker, the local artists, the people who work in the hospitality industry, or an eighty-year-old woman you meet at Piggly Wiggly.

There are smaller stores that have been in business for as long as I can remember, yet you will find state-of-the-art condominiums along the same stretch of beach. You'll see tacky buildings that have garish paint colors and beautiful Art Deco homes and townhouses, all combined into an eclectic mix of color and style.

You'll find some of the best Southern cooking from your family, friends, or even your next-door neighbor, and you can eat fresh seafood that was swimming in the Gulf of Mexico just hours before. And if you like both Southern cooking and seafood—I make an awesome shrimp and grits.

Speaking of Southern cooking and family, my parents always evacuated for hurricanes. My mom would gather up her prized possessions (her photo albums) and we'd get in the car and head to Dothan, Alabama. As I became an adult, we never had to evacuate because the storms that did affect us were nothing like this one. Sure, we rode out Hurricane Opal, but that was a category 3 and didn't hit us head on. Keep in mind that we have never had a storm like this in our area—ever.

People often wonder why people stay for a storm, but this one got strong fast, and yes, we were prepared with flashlights, candles, water, dog food, canned food, a battery powered radio, and all the usual things (except for Vienna Sausages because those are just gross). But no one knew it would turn into the powerhouse that it became.

Back when the storm was just a small one, we discussed with our neighbors how we'd all leave if it got too big, but at the time it was coming in as a category 2. Eventually, we were under an evacuation notice, but for the storm surge only, which never reaches my neighborhood. So again, many people stayed, including us.

By the time it was coming in stronger, it was too late to evacuate. We are surrounded by bridges, which close if the winds get up to a certain speed, and that can be as little as 45-50mph. We could have gone up Highway 231, which was congested with traffic while thousands of others headed out of town, but no one wants to get caught in a hurricane while on the highway. So along with our dogs, we hunkered down as they say, and prepared for the worst.

As it turns out, nothing I have ever experienced could have prepared me for Hurricane Michael. For hours, I lay in bed with my dogs and prayed. The winds were so strong I was scared to even look outside—and this comes from someone who loves storms, including the smaller hurricanes of the past. Usually you will find me taking some poorly made videos of bad weather but this one wasn't your usual storm.

I could hear large objects hitting the house and heard the trees falling on our neighbors' homes. In fact, I heard when our backyard neighbor's large sycamore tree hit their house. Fortunately, they yelled out a few curse words, not cries for help, so we knew they were okay. It was not as if we could have ventured out to help them anyway. I used to love listening to the breeze rustling the leaves on that tree. It's been around for decades and now it's gone.

As I lay in bed, I would press my hands against the walls and I swear to you, I could feel them "breathe." It was the most terrified I have ever been, and I thank God that I survived.

When the eye crossed over, I knew we had to go through hours more of the same thing. We'd lost power and phone service. There was nothing to do but wait, pray, and hope.

After the storm was over, we went outside and it was wonderful to see our neighbors out, some we had never even met, checking to see if anyone needed help.

Our close neighbors had a tree on their roof, along with their carport upside down on that same roof. And the church that has been in our neighborhood for decades was decimated. That is just some of the immediate damage we saw, and then I saw my own house.

The entire backyard was gone, the roof was severely damaged, we had water leaking into the ceiling, and the entire privacy fence was damaged or in some cases, missing entirely. The porch posts blew away, the satellite dish tore off part of the roof, and trees we'd had for years were twisted—that is, the ones that remained. All of the storage sheds were gone, with years of memories strewn throughout the yard along with tools, the Christmas tree and decorations, and ironically, a brand new generator.

After the storm, we sat all night listening to the radio about the devastation and hearing some of the stories from people who could call in, or the DJs who had weathered it out at the station. It was hot, it was dark, and it was scary, but just hearing a friendly voice made us feel that we weren't alone in the world.

One bright spot was going outside and seeing the Milky Way from the house, something we can never do because of the light pollution. The other thing you notice after the power goes out all over town is how quiet it is. I remember in *The Stand* by Stephen King, after the epidemic had affected most of the population, it was so quiet that Frannie could hear Harold using his manual typewriter from far away—it's almost that quiet, other than a stray vehicle or the sound of a generator.

We've since started picking up the pieces. We are extremely fortunate and although we have damage to our home, we know people who have lost everything.

I've seen the best come out in people and the worst, but I believe that these trials don't work that way—either you were a good person to start with or you weren't. A hurricane doesn't make you suddenly become a person who will try to cheat people on a roofing job and a natural disaster doesn't make you suddenly care about your neighbors and how they fared during the storm.

Overall though, I know our area is filled with good people, just like the heroes from all over the country who came to help. Whether they brought supplies, helped connect power lines, or worked clearing debris—it's not just a job. It's being there for those who need it.

The entire landscape has changed—we heard that in some places, ninety-five percent of the trees are gone. Buildings that have been around for 100 years are gone, places we've visited for years are closing because of the damage, and you can even get lost in your own neighborhood because nothing looks the same.

While this section of the Gulf Coast may never be the same, we will persevere and rebuild. My beloved Mexico Beach has been leveled, but I can only hope that it comes back even stronger. My hometown of Lynn Haven was one of the worst hit areas, yet I know that they are resilient and it will grow into what I once knew. Panama City, where I live, is in shambles, but again, I know that it will flourish and prosper eventually. We are strong and we will not let this storm define us.

Damaged sign

We have strength and will rebuild

ACKNOWLEDGMENTS

Special thanks to those who helped with some of the photos. Richard Hill, Bruce Benedict, Bill Fauth, Mike Cassini, Chastity Wofford, Lou Columbus, Joshua Pyfrom, and Laura Risner. Without them, I'd have less to share with everyone and lose out on a different perspective. Their keen eye allowed me to add even more to the book and I appreciate their help.

SOURCES AND END NOTES

1. Mazzei, Patricia. "Among the Ruins of Mexico Beach Stands One House, 'Built for the Big One,'" *New York Times,* October 14, 2018. https://www.nytimes.com/2018/10/14/us/hurricane-michael-florida-mexico-beach-house.html

2. https://en.wikipedia.org/wiki/Hurricane_Michael